# THE CHANNELED SCABLANDS OF EASTERN WASHINGTON

## —The Geologic Story of the Spokane Flood—

by
Paul Weis
and
William L. Newman

Presented to

## Okanagan College Library

by

*Nigel Skermer*

First Edition:   U.S. Department of the Interior, 1974
U.S. Government Printing Office: 1982—359-019

Second Edition:   Eastern Washington University Press, 1989, 1999
Cheney, Washington 99004

ISBN:   0-910055-11-4

**Fig. 1.**—Soil covered land sown with wheat is shown on the right; the channeled scablands are on the left.

Fig. 2.—Geographic setting of the Channeled Scablands, eastern Washington.

A traveler entering the State of Washington from the east crosses a flat-to-rolling countryside of deep, fertile soil commonly sown with wheat. Continuing westward, he abruptly enters a deeply scarred land of bare black rock cut by labyrinthine canyons and channels, plunge pools and rock basins, cascade and cataract ledges, displaying ragged buttes and cliffs, alcoves, immense gravel bars, and giant ripple marks. The traveler has reached the starkly scenic "Channeled Scablands," and this dramatic change in the landscape may well cause him to wonder, "What happened here?" The answer—the greatest floods yet documented by Man.

This publication summarizes the sequence of geologic events that culminated in the so-called Spokane Flood. It was prepared in response to the general interest in geology and the particular interest in the Channeled Scablands often expressed both by residents and by visitors to this part of the State of Washington.

# Geologic Setting

The name "Channeled Scablands" was first used in the early 1920's by geologist J. Harlen Bretz of the University of Chicago, whose comprehensive study of the region proposed the idea that the erosional features were the result of a gigantic flood. Although Bretz' concept sparked a lively controversy, geologists today agree that the Scablands were carved by a series of floods of unprecedented proportions during the last Great Ice Age.

The Spokane Flood left its mark along a course of more than 550 miles, extending from western Montana to the Pacific Ocean. The most spectacular flood features were carved into the black volcanic rock terrain in eastern Washington. This rock, the "floor" of the Scablands, is basalt—a dense crystalline lava that covers more than 100,000 square miles in parts of Washington, Oregon, and Idaho. The lava field that underlies the Scablands in eastern Washington is a saucer-shaped area of about 15,000 square miles almost

**Fig. 3.**—Chief Joseph Canyon in northeastern Oregon, is cut through many flows of Basalt.

Fig. 4.—Steptoe Butte, an island of older rock that stands above the "sea" of surrounding basalt. In this area, both it and the surrounding basalt lava are covered by a "frosting" of windblown dust called loess.

completely surrounded by mountains and nearly encircled by three rivers—the Columbia, the Spokane, and the Snake (Fig. 2, page 3).

Most of the lava flowed during the Miocene Epoch of the Tertiary Period, between about 25 million and 12 million years ago. At times, one flow followed another at short intervals, but at other times, tens of thousands of years intervened between flows. Erupting from long wide fissures, the molten rock flowed onto a hilly terrain of older rocks—an area that probably looked much like the present-day country north of the lava field. The molten material in the fissures crystallized to form dikes. In places, erosion has exposed swarms of dikes that cut older rocks. These places are believed to be major sources of the lava.

Early flows filled the valleys and subsequent flows covered most of the high hills, as layer upon layer of lava eventually formed a solid sea of basalt, in places more than 10,000 feet thick. Several individual flows, with thicknesses of more than 75 feet, have been traced for more than 100 miles.

Around the edges of the lava field a few hills poke up, island-like, through the basalt. One of the most prominent of these, Steptoe Butte, near Colfax, Washington, has given its name to all such features. Geologists call any island of older rock surrounded by lava a steptoe.

Fig. 5.—Columnar Basalt. Large regular columns exposed in a road cut near Spokane.

Molten lava occupies a greater volume than solidified lava. As fresh lava slowly cools and solidifies, a regular, prismatic pattern of shrinkage cracks commonly develops (Fig. 5). These joints break up the lava into vertical columns of rock. Many basalt flows in eastern Washington display this feature called columnar jointing.

Where basalt encroached upon a pond or lake, the quickly-quenched lava formed rounded blobs called "pillows" instead of columnar joints. In the western part of the lava field, a flow of basalt engulfed a swamp forest. Because of the water, the molten rock formed pillows and did not completely consume the surrounding plants. Among the fossilized remains of these plants are logs of the famous Ginkgo tree, a species that has survived for 250 million years. In 1934, an area of about 6,000 acres near the town of Vantage, Washington, where the vestiges of this ancient swamp are exposed, was set aside as the Ginkgo Petrified Forest State Park (Fig. 2, page 3).

Fig. 6.—Artist's concept of how Diceratherium might have appeared.

At a different time, but less than 50 miles northeast of the site of the swamp forest, another advancing lava flow encountered a shallow pond containing the floating body of a dead rhinoceros. Again, water prevented the molten rock from completely consuming the flesh; instead, pillows of lava enclosed the body forming a crude mold. In 1935, the site of the incident was discovered in the form of a cavity in a basalt cliff (Fig. 2, page 3; Fig. 6). The cavity contained several teeth and numerous pieces of fossilized bone. Studies of the shape of the cavity and the fossil remains showed the rhino to be one of an extinct species of Diceratherium.

After the eruptions ended, the lava field was tilted as a unit to the southwest. Today, the northeast rim of the saucer-like field is about 2,500 feet above the sea level, whereas its lowest point, near Pasco, Washington, is less than 400 feet above sea level. In addition to regional tilting, the lava field was deformed in places to yield a series of fold ridges. Saddle Mountain, Frenchman Hills, and Horse Heaven Hills are examples. Several of these can be seen along the west side of the lava field between Wenatchee and Pasco (Fig. 2, page 3). Of particular interest is the Coulee Monocline, an asymmetric fold that trends across the north-central part of the lava field (Fig. 7). The physical features of this monocline were to play an important role in determining the nature of the erosional features of Grand Coulee carved during the Great Flood.

Beginning some time after the flows of lava ended, a cover of windblown silt or loess began to accumulate over much of the lava field, eventually producing the fertile soils of the Palouse country of southeastern Washington. The loess attained its maximum thickness in the Pullman-Colfax area where it is locally as much as 200 feet thick and displays a distinctive rolling surface with steep north-facing slopes (Fig. 8).

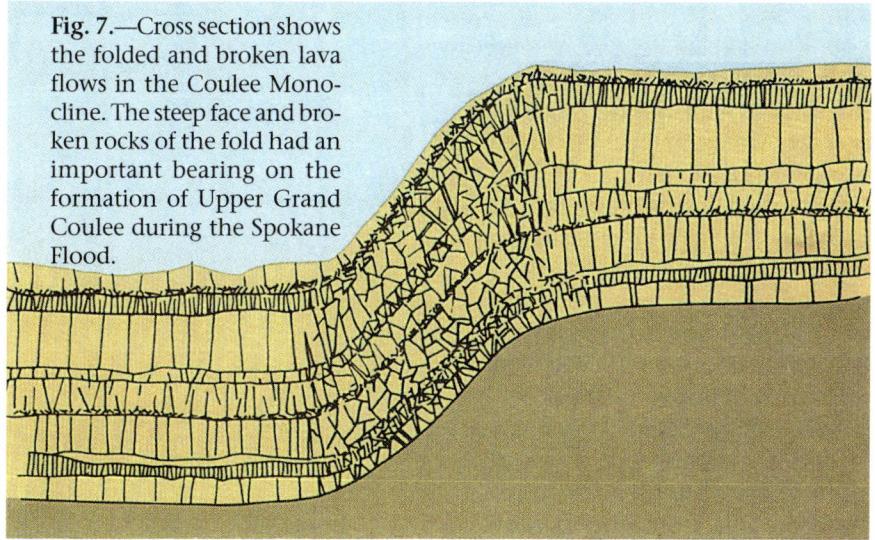

Fig. 7.—Cross section shows the folded and broken lava flows in the Coulee Monocline. The steep face and broken rocks of the fold had an important bearing on the formation of Upper Grand Coulee during the Spokane Flood.

**Fig. 8.**—Palouse Hills. Rolling hills of windblown silt (loess) in the rich wheat country of eastern Washington.

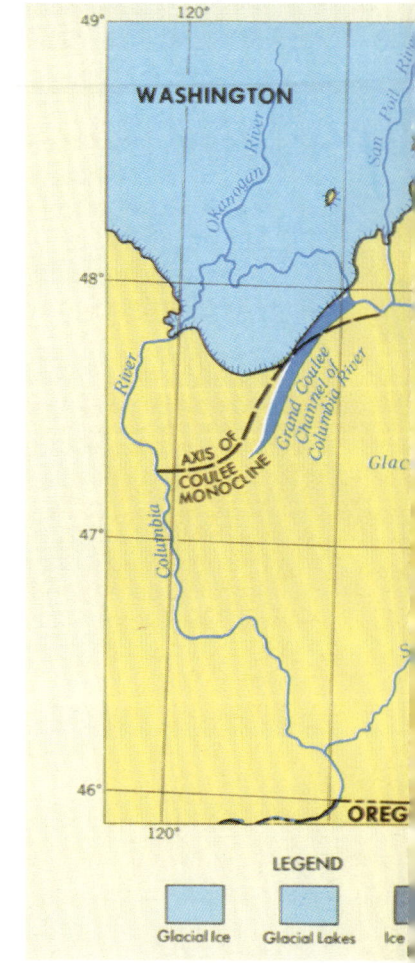

Some of the loess came from the western part of the lava field, where temporary lakes formed during the folding. The Palouse loess also contains volcanic ash, blown in from the volcanoes that are scattered along the Cascade Mountains of Washington and Oregon.

Thus, before the onset of the Great Flood, the geologic setting of the scabland region consisted of a thick, tilted saucer of basalt, in places warped into ridges and completely overlain by a "frosting" of loess. How did the region look? If one were standing atop Steptoe Butte a million years ago, the view in any direction would be peaceful rolling grassland speckled with roaming herds of antelope, buffalo, and camel with the distant mountains of the Cascades and Rockies forming hazy blue backdrops to the West and East. This tranquil scene was the setting for a catastrophe!

## The Great Ice Age

Although glaciation began in the northern hemisphere at least two million years ago, the glacial advances responsible for all the Scablands' floods are probably less than 750,000 years old, and by far the most evident of the floods are believed to have occurred 16,000 to 12,000 years ago. At that time, glaciers were moving southward from great ice fields in southern British Columbia, following major south-trending valleys leading into the United States. These valleys from west to east are Okanogan, San Poil, Columbia, Colville, Pend Oreille, and Priest River Valleys and the Purcell Trench, a great valley that in places contains the Kootenai River (Fig. 9).

The Okanogan Ice Lobe advanced far out into the lava field, in the process damming the Columbia River, forming a temporary lake in the

**Fig. 9.**—The advance of glacial ice and the corking of the Clark Fork River. Note that the formation of Glacial Lake Missoula required the existence of a large river system located in deep mountain valleys which lay almost entirely to the south of the glacial ice margins.

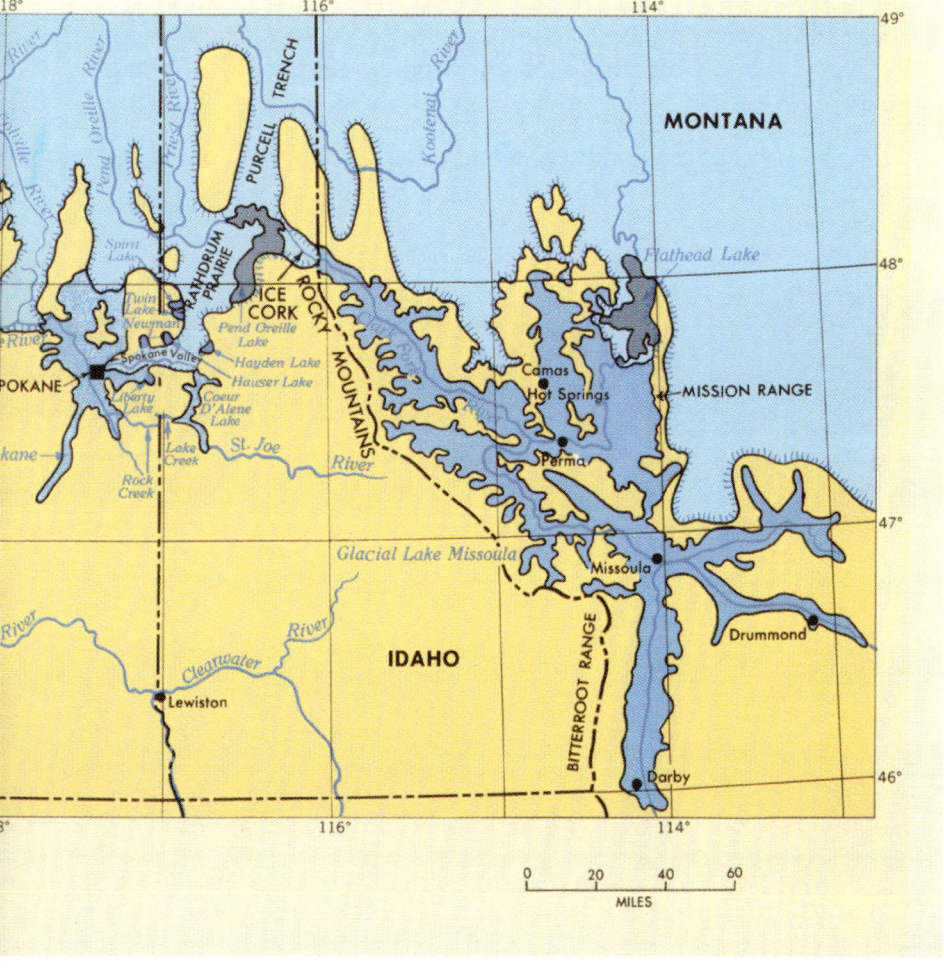

valley upstream that reached an elevation of about 2,300 feet above sea level. This diverted the Columbia River from its course and forced it to flow southward along the Coulee Monocline.

Farther east the Colville Lobe reached the edge of the lava field and dammed the Spokane River, creating Glacial Lake Spokane whose surface also reached an elevation of about 2,300 feet.

The Purcell Lobe scoured its way into a basin that now holds Pend Oreille Lake, and from there moved southwestward across the Rathdrum Prairie and the Spokane Valley to a point very near the eastern city limits of Spokane.

The glacier deposited a mixture of rocks, sand, and dirt along its margins, forming lateral moraines and kame terraces. This debris dammed the mouths of side valleys along the glacier's path and created many lakes. Spirit Lake, Twin Lakes, Hauser Lake, and Newman Lake on the northwest side of the glacier near Spokane are remnants of larger lakes that existed while the ice occupied the main part of the valley. On the southeast, Hayden Lake, Coeur d'Alene Lake, and Liberty Lake are other examples of lakes formed when the stream valleys became partly or entirely blocked by dams of glacial debris.

The largest lake that formed in the Pacific Northwest during the Great Ice Age was Glacial Lake Missoula, the lake that spawned the flood (Fig. 9, pages 8 & 9). We know now that the lake formed and emptied dozens of times. Each breaching of the ice dam produced an enormous flood, but it still ap-

**Fig. 10.**—Cliff buttress at northern end of Bitterroot Range and mouth of Clark Fork River—the site of the ice dam that formed Glacial Lake Missoula.

**Fig. 11.**—Wave-cut shorelines of Glacial Lake Missoula are shown on Sentinel Mountain, Missoula, Montana. The lake once had a depth of 950 feet where the University of Montana stadium now stands.

pears that the largest, deepest lake, which produced the greatest of all the floods, was one of the last to occur.

## Glacial Lake Missoula

As the Purcell Lobe moved southward within the Purcell Trench, glacial ice plugged the Clark Fork Valley like a giant cork, with the 2,000-foot cliff at the extreme northern end of the Bitterroot Range serving as a buttress. The ice dammed the Clark Fork River near the point where it empties into Pend Oreille Lake.

The water that was impounded behind the dam filled the tributary valleys for many miles to the east. At its highest level, Glacial Lake Missoula covered an area of about 3,000 miles and contained an estimated 500 cubic miles of water—half the volume of present day Lake Michigan. Its surface stood at 4,150 feet above sea level, giving the lake a depth of nearly 2,000 feet at the ice dam (more than twice the depth of Lake Superior). Traces of ancient shorelines in western Montana indicate that the lake was about 950 feet deep at present-day Missoula, 260 feet deep at Darby, and more than 1,100 feet deep at the south end of Flathead Lake. The glacial lake's eastern shore was someplace east of Drummond. The Drummond area itself was beneath nearly 200 feet of water.

The shorelines of Glacial Lake Missoula are still visible in several place and are especially well shown in Missoula (Fig. 11).

The wave-cut shorelines are faint and delicate, suggesting that no one stand of the lake was of long duration. The close spacing of the shorelines may indicate successive water levels as the lake gradually filled.

While ice plugged the mouth of the Clark Fork River at the Idaho-Montana border, other glaciers contributed meltwater to the growing lake. A large lobe of ice from Canada reached the south end of present-day Flathead Lake, and at this point its rate of melting was greater than its rate of advance. Its terminus became part of the north shore of Glacial Lake Missoula, and its meltwater poured into the lake. Alpine glaciers in the Bitterroot Range, the Idaho Rockies, and the Mission Range added their meltwater, and summer rains also helped raise the level of the lake.

Eventually, the lake level reached the lip of the ice dam and began to overflow. One can only guess at the details from this point until the dam was breached completely, but some aspects of the event can be surmised. The overflowing stream must have cut deeply into the ice. As the outlet was quickly lowered, the water roared through the breach, probably widening the channel rapidly by undercutting the sides until the dam collapsed. Huge icebergs must have been ferried along by the stream. Within a very short time—perhaps no more than a day or two—the ice dam was destroyed and the contents of the lake were released.

## The Spokane Flood

When the ice dam at the mouth of the Clark Fork River failed, the lake drained at a rate unmatched by any flood known, and the water had only one place to go—south and southwestward across Rathdrum Prairie and down the Spokane Valley (Fig. 12), which by this time was probably ice free. As the lake basin drained, the water had to pass through narrow parts of the Clark Fork Canyon where current velocities are calculated to have reached 45 miles per hour. The maximum rate of flow here has been estimated to have reached 9 ½

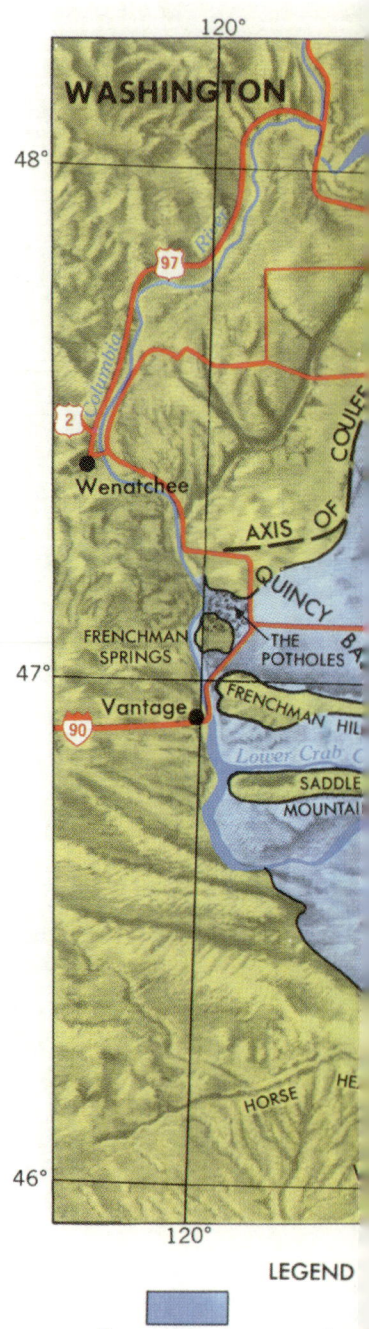

Fig. 12.—The Channeled S

12

**Fig. 13.**—Giant ripple marks. Long gravel ridges formed by the powerful currents that flowed across Markle Pass, near Camas Hot Springs, Montana. The view is toward the northwest.

to 16 cubic miles per hour, or at least 400 million cubic feet per second (c.f.s.). Nine and one-half cubic miles per hour is roughly equivalent to ten times the combined flow of all the rivers of the world. For comparison, the rate of flow of the world's largest river, the Amazon, is 6 million cubic feet per second. The Columbia River at The Dalles, Oregon, averages 195,500 c.f.s., and the Spokane River at Spokane Falls averages about 6800 c.f.s.

For decades, no one knew how may times the Ice Age Floods ravaged the Columbia Plateau. It was not until the 1980's that Brian Atwater of the U.S. Geological Survey found at least a large part of the answer. He studied the deposits in the San Poil River estuary, about 50 miles east of Grand Coulee. The reason: it was the only significant stream within the flood's path that enters the Columbia River from the north. Whenever Glacial Lake Missoula emptied, the resulting flood scoured or reworked everything in its path. But they merely filled the lower reaches of the San Poil, dumping flood debris on top of the fine-grained, banded, annual silts that were the normal river sediments. When no great floods occurred, those fine sediments were the only deposits to form. Atwater was able to show evidence of 89 great floods.

As the vast basin of Lake Missoula drained, local deposits were formed that bear striking testimony to the power of the moving water. Perhaps the most dramatic of these are the "giant ripple marks" that formed in many places. These gravel ridges are plainly visible on aerial photographs, but went unnoticed for many years simply because their immense size makes their pattern and symmetry almost indistinguishable from the ground. A series of these ripple marks is particularly well displayed on the south side of Markle Pass, where they are cut by Montana State Highway 28 between Perma and Camas Hot Springs (Fig. 13).

Water poured across the ridge from the broad valley north of the pass in a cascade at least 800 feet deep, scoured the rocks at the ridge crest, and plucked away at the bedrock to form the two basins that now contain the Schmitz Lakes. South of the pass, the currents deposited ridge after ridge of coarse gravel in great sweeping curves. These ridges are typical ripple marks in every respect except for their great size. They range up to 35 feet high, with crests more than 350 feet apart. The largest are at the north. Some of the ridges are up to 2 miles long, and altogether they cover an area of about 6 square miles.

The powerful, turbulent currents of the flood moved material of all sizes, including huge boulders (Fig. 21). One characteristic of the flood boulders is

**Fig. 14.**—Thin, fine-grained sedimentary layers in this exposure were deposited annually in the San Poil River estuary during the latter part of the last glacial period. The much thicker, darker, poorly sorted layers are Ice Age Flood deposits. Eighty nine such layers have been identified in the Sanpoil watershed.

Fig. 15.—Streamlined, teardrop-shaped flood gravel bar deposited on the down-stream side of a bedrock barrier near Shelly Lake, south of Veradale, Washington.

the presence of percussion marks—chips and nicks made as the boulders collided with one another while suspended in the flood. Sediments ranging in size from clay particles to boulders were heaped together where the currents dropped them to form deposits of chaotic debris.

The water that poured out of the mouth of the Clark Fork Valley swept through Pend Oreille Lake, spread across the Rathdrum Prairie, and roared down the Spokane Valley. An arm of the flood surged up Coeur d'Alene Lake and spilled across the divide between Lake Creek and Rock Creek, but most of the water flowed down the Spokane Valley to the north rim of the lava field.

## The Carving of the Scablands

As noted earlier, the lava field is shaped somewhat like a giant saucer, tilted to the southwest. Slopes (gradients) down the tilted northern rim are as much as 25 to 35 feet per mile (for comparison, the Spokane River between Coeur d'Alene and Spokane has an average gradient of only 9 feet per mile). When the flood reached the lava field and started down this sloping surface, the enormous volume, velocity, and turbulence of the water provided the erosional energy required to sweep away the loess and expose the jointed basalt underneath. The currents were so turbulent and so powerful that they

**Fig. 16.**—Palouse "island"--an erosion remnant of Palouse soil, along Interstate Highway 90 about 5 miles north of Sprague.

were able to pluck out and transport blocks of basalt, some measuring more than 30 feet across. Deep canyons were eroded into the basalt and, where cascades developed, plunge pools and cataracts formed. Some of these depressions are more than 200 feet deep. Many of the lake basins in the northeastern part of the area were formed in this manner. In other places, great amounts of stream-borne rock debris were dumped to form immense gravel bars and deltas.

Three giant rivers raced across the lava field (Fig. 12, page 12). The easternmost stream, 20 miles wide in places and locally 600 feet deep, carved the widest channel, the Cheney-Palouse Tract. A middle river carved the Crab Creek Channel and its tributaries, a tract about 14 miles wide crossed near its head by U.S. Highway 2 between Davenport and Creston. The third and westernmost river, which may have carried the greatest volume of water, carved Grand Coulee.

These three major torrents, together with dozens of smaller streams, flowed simultaneously and at times criss-crossed their channels until the 500 cubic miles of water stored in Glacial Lake Missoula had drained away (inside back cover).

At Sprague, in part of the Cheney-Palouse Tract, flood waters swirled along a channel eight miles wide and more than 200 feet deep. Here it stripped

**Fig. 17.**—Upper Grand Coulee, looking south. Steamboat Rock is a lava remnant that forms the flat-topped island-like mesa in the Coulee. In front of it are exposures of granite that form the floor of the upper end of the Coulee.

away 100 to 150 feet of Palouse soil and parts of jointed lava flows, leaving a wild rough wasteland in its wake. This jumble of mesas and depressions is so irregular that an area of more than 70 square miles contains no throughgoing streams.

The sides of the scablands channels in the eastern part of the lava field are marked in many places by steep slopes cut in the Palouse soil revealing in spectacular fashion the depth of the soil. South of Ewan, Washington, for example, eroded scarps are nearly 200 feet high (Fig. 1, page 2).

The largest of the scablands channels is Grand Coulee, a two-stage canyon 50 miles long and as much as 900 feet deep. Its ancestral channel was cut by the Columbia River when it was forced to flow across the lava field by the Okanogan Ice Lobe. When flood waters cascaded down the steep southeast-facing slope of the monocline, this

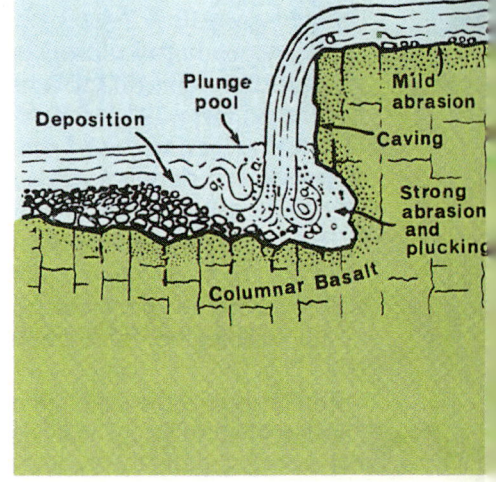

**Fig. 18.**—Cross section illustrating formation of falls, undercutting, plunge pool, and cataract retreat. Upper Grand Coulee was eroded by the process of cataract retreat. Water in the plunge pool at the base of the falls undercuts the rock on the upstream side, causing the lip of the falls to collapse and forming a new lip further upstream.

fold was eroded upstream across the zone of broken rocks and into a higher, flat-lying basalt, forming a typical recessional gorge now called Upper Grand Coulee (Fig. 18). Erosion and cataract retreat continued in Upper Grand Coulee until the basalt flows at the head of the Coulee were completely removed. This exposed the granitic rocks that form the Coulee floor and the foundations for Grand Coulee Dam today. Like the basalt, the granite is hard and tough, but unlike the basalt, the joints in the granite are widely spaced and

**Fig.19.**—Aerial view of Dry Falls. Banks Lake (Upper Grand Coulee) is in the upper right.

irregular, and it therefore was not plucked and eroded by the flood. But erosion did lower the divide between the Columbia and the Coulee to a point where the falls essentially destroyed itself.

Lower Grand Coulee, cut along the foot of the monocline, also developed a series of falls and plunge pools as erosion progressed upstream (Fig. 19). Dry Falls, a 350-foot-high, 3-mile-wide group of scalloped cliffs shows today the position of the falls at the end of the flood.

South of Upper Grand Coulee, the water from the Coulee channel was joined by the great river that boiled down Crab Creek. Together they surged into Quincy Basin, spread, slowed, and dumped a great blanket of boulders, sand, and gravel—material mostly derived from flood erosion upstream. (Fig. 21). This flood debris now covers an area of about 500 square miles to a depth of at least 125 feet. As Quincy Basin filled, some of the water overflowed through two large cataracts into the Columbia River Valley at the Potholes and Frenchman Springs dischargeways (Fig. 12, page 12). But most of the water escaped to the south around the east end of Frenchman Hills.

At the east end of Frenchman Hills, the flood crossed a 50 square mile area covered by a layer of weak sedimentary rocks called the Ringold Formation

Fig. 20.—Flood-eroded Crab Creek channel near Odessa, Washington.

which had been deposited during early Pleistocene time (2 to 3 million years ago). This part of the Scablands, called the Drumheller Channels, is perhaps the most spectacularly eroded area of its size in the region (Fig. 12, page 12). The erosional features can be seen especially well from the road built on the crest of O'Sullivan Dam, along the north edge of the Channels (Washington State Highway 262). South of the dam is a wild jumble of cliffs, depressions, ponds, and remnants of lava flows. This channel area has a local relief of 150-200 feet.

South of Frenchman Hills, another branch of the flood flowed westward in Lower Crab Creek Valley along Saddle Mountain, and near here this branch was joined by more flood waters from Washtucna and Lind Coulees. These surging torrents cut the Othello Channels at the east end of Saddle Mountain (Fig. 12, page 12).

Like the proverbial roads to Rome, all of the scabland rivers led to Pasco Basin, a down-warped area of more than 500 square miles at the southwest corner of the lava field. The southwestern edge of Pasco Basin is marked by the Horse Heaven Hills, which was an effective barrier to the escape of water except for a single outlet—Wallula Gap (Fig. 22, page 22). Because all flood waters from the lava field had to pass through this narrow squeezeway, a huge lake formed in Pasco Basin and extended more than 100 miles upstream in the valley of the Snake River. At the junction of the Snake and Clearwater Rivers at Lewiston, Idaho, this impounded water was almost 600 feet deep (Fig. 12, page 12).

Leaving the Gap, the flood waters moved westward into the Columbia River Gorge, through the Cascade Mountains into the Portland Basin. There

**Fig. 21.**—Flood boulder in the Quincy Basin, about three miles south of Ephrata, Washington. It measures 59 x 36 x 26 feet. It is the largest of thousands of boulders flushed out of Lower Grand Coulee by Ice Age Floods.

**Fig. 22.**—Wallula Gap, looking northeast (upstream) to the valley cut through Horse to the top, and stripped the soil from the basalt on both sides.

the great stream created a large delta and continued on up the Willamette Valley to form a short-lived lake nearly 400 feet deep. As this lake gradually receded, huge icebergs loaded with rock debris were stranded along its shoreline. When the ice melted, the debris, containing boulders as much as seven feet across, was left behind as mute evidence of the immense size and power of the flood.

Did early man witness the fearful destruction of the land by the raging torrent? At this point, no one knows for sure, because the earliest evidence of man in the region has been dated at about 10,130 years B.P. (before the present), or a few thousand years after the flood, as determined in the Carbon-14 laboratory of the U.S. Geological Survey from charcoal collected at the 'Marmes Man' site in southeastern Washington (Fig. 12, page 12). This prehistoric rock shelter, carved along the base of a basalt cliff by the Palouse River, is believed to be among the oldest known inhabited sites in the United States. Thus, at the present time, definite proof is lacking that man was living in the pacific Northwest at the time of the flood.

The duration of the flood is not known, but a reasonable guess is that at any one point on the lava field it consisted of a series of surges, and that the crest was short-lived—perhaps lasting only a day or two. At Wallula Gap,

Heaven Hills by the Columbia River. When the flood occurred, water filled this canyon where the maximum flow was calculated at nearly 40 cubic miles per day, the main part of the flood probably lasted 2 to 3 weeks.

The maximum rate of flow at the ice dam has been estimated at 9.5 cubic miles of water per hour. If this rate were sustained, Glacial Lake Missoula would have drained in little more than two days. The maximum rate of flow at Wallula Gap has been estimated at 39.5 cubic miles per day, which would drain the lava field in about 2 weeks. But these rates probably could not have been maintained for that long. A realistic estimate of the duration of the flood, from the time when water first spilled over the ice dam to the time when scabland streams returned to nearly normal flow, may have been nearly a month.

The unique combination of geologic events, beginning with a vast series of lava flows, then regional tilting of the land, followed by deposition of 100-200 foot layers of windblown silt, and ending with glacial lakes repeatedly formed and released to form the earth's greatest known floods, involved such a large area that only parts of the scabland picture can be seen at one time. Now, as a result of the detailed studies of many geologists working in the area for more than fifty years, the many fragments of evidence have been pieced together to further confirm Bretz' concept of the Great Spokane Flood.

**Fig. 23.**—This unique view of the Channeled Scablands region, covering an area of about 13,225 square miles, was taken on August 31, 1972 by the first NASA Earth Resources Technology Satellite [ (ERTS-1) now called Landsat] from an altitude of 569 miles.

The dark 'braided' pattern clearly depicts the channelways of the Great Spokane Floods—the areas where vigorous stream erosion stripped away the "frosting" of loess to expose underlying dark basalt.

A large part of the region is planted with wheat as the checkered appearance of the terrain denotes. The clusters of small red circles at the lower left are fields irrigated with rotating sprinklers as part of the Columbia Basin Reclamation Project. The dark red area to the north of the Columbia and Spokane Rivers is the densely timbered region called the Okanogan Highlands. Grand Coulee and Grand Coulee Dam are at the upper left. The stream in the lower right of the picture is the Snake River.

* * *

This color "photograph" was prepared from a set of three Earth images recorded in the green, red, and infrared bands of the spectrum by the Landsat Multispectral Scanner. The electronic data was transmitted to a ground station and processed. The three images were combined, using appropriate color filters, to make this "false" color composite. Green vegetation appears red on this composite mainly because of the strong response of the infrared band to chlorophyll.

Each of the spectral bands tends to enhance certain ground conditions; study of the various images, separately or combined, leads to a better understanding of the nature of the Earth's surface. Suitable coverage of the entire United States is being acquired. The satellite provides images of the same area every 18 days enabling scientists to monitor natural and man-made changes in the Earth's surface.

* * *

Inquiries concerning Earth imagery should be addressed to the EROS Data Center, U.S. Geological Survey, Sioux Falls, South Dakota 57198